BRAMWELL's SUBTRACTION METHOD (BSM)

(The fun and easy way to do subtraction)

By

DENNIS L BRAMWELL

BSM press.

ISBN : 978-0-9926897-0-4

50% of all proceeds to be donated to charity.

Published by BSM press

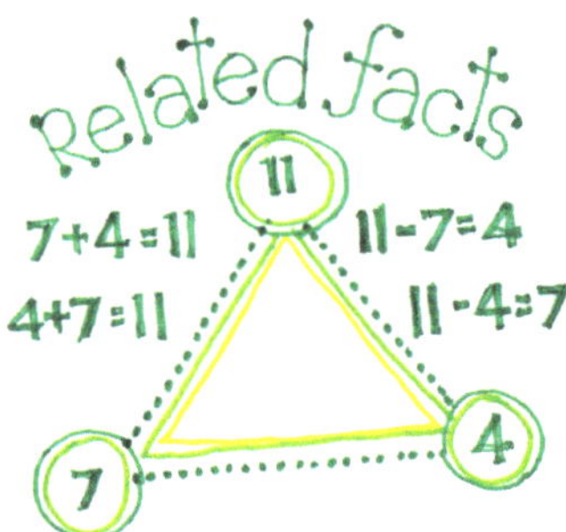

Book design by Donya Davis

Illustrations by Mike and Paula Conrad

Cover design by Rita the dog
(http://rita314.wordpress.com/)

Please, I would like you to study and understand this new and fun method of Subtraction (do not worry, it is very easy).

You will then be able to help your child if ever s/he makes a mistake.As you go through this work book with your child, you may either mark their work or else you can let them check their answers against those at the back of the book.

Be indulgent toward their mistakes, gently correcting them. Mistakes mean that the child did not fully understand the procedure. Please point out where they went wrong. If necessary, give them the same exercise again until they get 90% or 100%.

It is exceedingly motivating for a child to get 90% or 100%. It greatly encourages a child to strive for excellence. Often this can lead to an increased love of Mathematics.

By working through this work book with your child, s/he will quite likely start improving in all his or her mathematical topics.

The Beauty of BSM

BSM is easier to master than the traditional subtraction methods, and experience with teaching BSM shows that it is mastered more quickly. With BSM there is no taxing mental arithmetic because you never have to subtract from a number greater than 10 – an ability that children always pick up very quickly with the assistance of using their fingers as and when necessary.

Compare this with mentally subtracting from numbers larger than 10 (e.g.14 – 6), a requirement when using any of the traditional subtraction methods and you will see that getting an answer (for most people) will not come quite so immediately.

And when the taxing bits are taken out of subtraction, the whole process becomes less effort-laden, and with the surplus effort removed, BSM delivers the following benefits:

The task of subtraction is made much easier.

●

It's much harder to make a mistake.

●

Success with subtraction
becomes the norm

●

The confidence of all children will be boosted by repeated
quick and unpressured success because all children will
end up seeing themselves as "maths capable" rather than
"maths incapable".

●

Many more children will find that maths can be fun – as
opposed to being a source of frustration.

●

Many more children will show an interest in mastering
further maths skills, one step at a time.

●

Learning and using BSM is fun and satisfying for all
children.

●

Teaching BSM is satisfying and fun for us teachers, as we
can get delight from seeing our students blossom in front
of our eyes.

CONTENTS

Warm up exercises — 6

Easy subtraction sums — 8

Bramwell's Subtraction Method (BSM) — 10

Starting BSM in the middle of a sum — 17

Coming out of BSM in the middle of a sum — 20

Going in and out and in again and out — 22

An alternative to going in and out and in — 28

Word Problems — 32

Finding missing digits (Optional) — 36

How to adapt BSM for mental calculations — 38

Appendix 1
How BSM came to be discovered — 48

Appendix 2
Traditional methods of Subtraction — 50

Appendix 3
Why BSM works — 58

Answers — 60

In this book, you will learn how to do subtraction sums in a very simple way. To begin, let us do some warm up exercises with simple numbers.

Practise the examples on the next two pages
(recommended time : 5 minutes)

9 – 1 = ___	8 – 2 = ___	4 – 3 = ___
9 – 2 = ___	8 – 4 = ___	4 – 2 = ___
9 – 4 = ___	8 – 6 = ___	3 – 2 = ___
9 – 3 = ___	8 – 5 = ___	3 – 1 = ___
9 – 5 = ___	7 – 1 = ___	2 + 7 = ___
9 – 7 = ___	7 – 4 = ___	2 + 6 = ___
9 – 6 = ___	7 – 2 = ___	2 + 4 = ___
9 – 8 = ___	7 – 6 = ___	1 + 5 = ___
9 – 9 = ___	7 – 5 = ___	2 + 3 = ___

10 – 1 =		7 – 3 =		2 + 2 =	
10 – 2 =		6 – 1 =		3 + 3 =	
10 – 4 =		6 – 4 =		3 + 6 =	
10 – 3 =		6 – 2 =		3 + 5 =	
10 – 5 =		6 – 5 =		3 + 2 =	
10 – 7 =		6 – 3 =		4 + 4 =	
10 – 6 =		5 – 3 =		4 + 5 =	
10 – 8 =		5 – 2 =		6 + 1 =	
10 – 9 =		5 – 1 =		7 + 1 =	
8 – 1 =		5 – 4 =		3 + 4 =	
8 – 3 =		4 – 1 =		2 + 5 =	

60

In the simple sums above the child is allowed no more than 2 mistakes. This equates to getting 97% or more. If more than 2 mistakes are made, they must be corrected and the child must repeat all 60 sums at a future date until they make no more than 2 mistakes. They must not be allowed to continue with this book if they persist in making more than 2 mistakes.

In a subtraction sum like :

$$
\begin{array}{r}
7\ 8\ 5 \\
-\ 2\ 4\ 3 \\
\hline
\end{array}
$$

Where the top digits are all greater than the corresponding bottom digits, we simply subtract in each column to get the answer, so :

$$
\begin{array}{r}
7\ 8\ 5 \\
-\ 2\ 4\ 3 \\
\hline
5\ 4\ 2
\end{array}
$$

These kind of subtraction sums are called easy subtraction sums.

1.
$$\begin{array}{r} 5\ 6\ 2 \\ -\ 3\ 2\ 1 \\ \hline \end{array}$$

2.
$$\begin{array}{r} 3\ 3\ 2 \\ -\ 1\ 0\ 1 \\ \hline \end{array}$$

3.
$$\begin{array}{r} 7\ 4\ 2 \\ -\ \ \ 3\ 1 \\ \hline \end{array}$$

4.
$$\begin{array}{r} 2\ 8\ 8 \\ -\ \ \ 4\ 7 \\ \hline \end{array}$$

5.
$$\begin{array}{r} 8\ 9\ 9 \\ -\ 4\ 0\ 3 \\ \hline \end{array}$$

6.
$$\begin{array}{r} 1,\ 7\ 4\ 2 \\ -\ \ \ \ 3\ 1\ 1 \\ \hline \end{array}$$

7.
$$\begin{array}{r} 8,\ 9\ 2\ 6 \\ -\ 3,\ 6\ 0\ 1 \\ \hline \end{array}$$

8.
$$\begin{array}{r} 7,\ 1\ 3\ 8 \\ -\ 6,\ 0\ 1\ 7 \\ \hline \end{array}$$

9.
$$\begin{array}{r} 2,\ 1\ 0\ 4 \\ -\ \ \ \ 1\ 0\ 2 \\ \hline \end{array}$$

10.
$$\begin{array}{r} 9,\ 9\ 9\ 9 \\ -\ 4,\ 5\ 3\ 6 \\ \hline \end{array}$$

In Exercise 1, there are 35 separate subtractions, the child is allowed at most 2 mistakes. If more than 2 mistakes are made, they must be corrected and all 35 subtractions must be repeated at a future date until they make 2 mistakes or less

35

Consider:

$$704$$
$$-389$$

This is where the top digits are less than the bottom digits in the units and tens columns.

In the units column, we cannot take 9 from 4.

There are various methods used to deal with this situation. The first method is a new method called Bramwell's Subtraction Method. (BSM)

The other more traditional methods will be discussed in Appendix 2.

You must remember the rule with the rhyme

"FIRST FROM 10, THE REST FROM 9, DROP 1 AT THE END OF THE LINE"

This is Bramwell's Rule

IT IS VITAL THAT THE PUPIL LEARNS THIS RULE OFF BY HEART BEFORE S/HE PROCEEDS.

FIND:
$$\begin{array}{r} 6,246 \\ -\ 2,378 \\ \hline \end{array}$$

START IN THE UNITS COLUMN	
Take **8** from **10** = 2 and **add 6** to give **8**	$\begin{array}{r} 6,246 \\ -\ 2,378 \\ \hline 8 \end{array}$
Take **7** from **9** = 2 and **add 4** to give **6**	$\begin{array}{r} 6,246 \\ -\ 2,378 \\ \hline 6\ 8 \end{array}$
Take **3** from **9** = 6 and **add 2** to give **8**	$\begin{array}{r} 6,246 \\ -\ 2,378 \\ \hline 8\ 6\ 8 \end{array}$
Finally Take **2** from **6** = 4 **drop 1** to get **3**	$\begin{array}{r} 6,246 \\ -\ 2,378 \\ \hline 3,868 \end{array}$

This is the procedure
Start in the units column
Subtract the 1st digit from 10 then ADD the top digit
Subtract the 2nd digit (tens digit) from 9 then ADD the top digit
Subtract the 3rd digit (hundreds digit) from 9 then ADD the top digit
Subtract in the last column in the normal way but you must write 1 less

FIND: 30,705
       −   827

START IN THE UNITS COLUMN
Take **7** from **10** =3 and **ADD 5** to get **8**

```
  3 0, 7 0 5
−       8 2 7
            8
```

Take **2** from **9** = 7 and **ADD 0** to get **7**

```
  3 0, 7 0 5
        8 2 7
          7 8
```

Take **8** from **9** = 1 and **ADD 7** to get **8**

```
  3 0, 7 0 5
−       8 2 7
        8 7 8
```

We are now at the end of the line
So we must **drop 1**,   30 − 1 = 29

```
  3 0, 7 0 5
−       8 2 7
  2 9, 8 7 8
```

FIND: 1 7
         −   8

Take **8** from **10** = 2 and **ADD 7** to get 9

        1 7
      −   8
      ———
          9

We are now at the end of the line
So we drop 1 (1 − 1 = 0) and we have the answer

        1 7
      −   8
      ———
          9

Example 

Find: 1, 0 0 0
         −   3 7 4

START IN THE UNITS COLUMN

Take **4** from **10** = 6 and **ADD 0** to get 6

      1, 0 0 0
    −   3 7 4
    —————————
            6

Take **7** from **9** = 2 and **ADD 0** to get 2

      1, 0 0 0
    −   3 7 4
    —————————
          2 6

Take **3** from **9** = 6 and **ADD 0** to get 6

      1, 0 0 0
    −   3 7 4
    —————————
        6 2 6

We are at the end of the line, so we drop 1
1 − 1 = 0 ; **so the answer is 6 2 6.**

      1, 0 0 0
    −   3 7 4
    —————————
        6 2 6

Using Bramwell's Subtraction Method, you only have to subtract from 9 or 10, then you have to add single digits whose sum is always a single digit.

This makes the work easy. Don't forget to drop one at the end.

Do the following subtractions using Bramwell's rule
**"First from 10, the rest from 9,
drop 1 at the end of the line "**.

1	4 3 4 − 1 7 6	6	6,7 1 4 − 5 6
2	3,0 0 0 − 1,8 4 2	7	7,2 1 2 − 7 5
3	4,2 1 1 − 2,9 8 9	8	20,1 4 3 − 7 8 5
4	2 6,2 3 1 − 1 9,8 6 7	9	3 0,0 1 2 − 7 4 6
5	3,5 2 1 − 4 9	10	4 0,1 2 3 − 2 8 7

In Exercise 2A, it is recommended that, as each question
is attempted, it is checked immediately.

If an error is made, the error must be pointed out to the
child and the procedure explained again, before the child
is allowed to attempt the next question.

1	1 5 − 7	6	1, 9 1 3 − 3 8
2	1 7 − 9	7	1, 0 0 0 − 3 2 8
3	2 5 − 8	8	5, 0 0 0 − 2, 7 6 2
4	3 0 0, 6 3 2 − 1 6, 8 5 7	9	8, 0 0 1 − 3, 7 4 6
5	8, 5 0 2 − 1 8	10	£ 5 0 . 0 0 − £ 2 7 . 6 2

In Exercise 2B, there are 32 separate subtractions, 2 mistakes equates to 93%. If a mistake is made, it must be explained to the child why the mistake was made, and corrected.

If more than 2 mistakes are made, the pupil must repeat the exercise again at a future date until they make 2 mistakes or less.

32

Sometimes we start using the new rule in the middle of a sum.

Example

FIND:
$$\begin{array}{r} 74{,}086 \\ -\,47{,}325 \end{array}$$

6 – 5 = 1

$$\begin{array}{r} 74{,}086 \\ -\,47{,}325 \\ \hline 1 \end{array}$$

8 – 2 = 6

$$\begin{array}{r} 74{,}086 \\ -\,47{,}325 \\ \hline 6\ 1 \end{array}$$

Now, 3 is greater than 0, so we start using the new rule.
We take **3** from **10** = 7 and **ADD 0** to get **7**

$$\begin{array}{r} 74{,}086 \\ -\,47{,}325 \\ \hline 7\ 6\ 1 \end{array}$$

7 from 9 is 2 add 4 = 6

$$\begin{array}{r} 74{,}086 \\ -\,47{,}325 \\ \hline 6{,}761 \end{array}$$

Finally 7 – 4 = 3 , drop 1 to get 2

$$\begin{array}{r} 74{,}086 \\ -\,47{,}325 \\ \hline 26{,}761 \end{array}$$

1	32,457 − 28,624	6	728 − 477	11	347 − 253
2	55,019 − 27,451	7	347 − 162	12	40,375 − 592
3	43,613 − 4,690	8	654 − 81	13	10,237 − 457
4	6,328 − 2,697	9	7,318 − 4,796	14	9,258 − 3,735
5	806 − 472	10	4,647 − 1,942	15	100,000 − 5,270

60

There are 60 separate subtractions in Exercise 3. If a mistake is made, it should be pointed out to the pupil why they went wrong.
No more than 5 mistakes are allowed. Allowing for no more than 5 mistakes equates to 92% or more. If more than 5 mistakes are made, the pupil must repeat the whole exercise, at a future date until they make 5 mistakes or less.

Example 

FIND:

$$\begin{array}{r} 6\ 8,\ 7\ 3\ 2 \\ -\ 4\ 3,\ 2\ 7\ 5 \end{array}$$

5 is greater than 2
so we start using BSM

5 from **10** is 5, **ADD 2** to get **7**

$$\begin{array}{r} 6\ 8,\ 7\ 3\ 2 \\ -\ 4\ 3,\ 2\ 7\ 5 \\ \hline 7 \end{array}$$

7 from **9** is 2 ,**ADD 3** to get **5**

$$\begin{array}{r} 6\ 8,\ 7\ 3\ 2 \\ -\ 4\ 3,\ 2\ 7\ 5 \\ \hline 5\ 7 \end{array}$$

Now 7 is greater than 2 so we are at the
end of the line and we drop 1.
7 − 2 = 5, drop 1, to get 4

$$\begin{array}{r} 6\ 8,\ 7\ 3\ 2 \\ -\ 4\ 3,\ 2\ 7\ 5 \\ \hline 4\ 5\ 7 \end{array}$$

8 is greater than 3, so we simply subtract
8 − 3 = 5

$$\begin{array}{r} 6\ 8,\ 7\ 3\ 2 \\ -\ 4\ 3,\ 2\ 7\ 5 \\ \hline 5,\ 4\ 5\ 7 \end{array}$$

6 − 4 = 2

$$\begin{array}{r} 6\ 8,\ 7\ 3\ 2 \\ -\ 4\ 3,\ 2\ 7\ 5 \\ \hline 2\ 5,\ 4\ 5\ 7 \end{array}$$

7 5 , 0 1 2
− 5 2 , 8 3 6
	6 4 7 2 , 2 7 7
	    − 2 1 0 , 4 9 1
		11 3 8 , 7 2 3
		    − 1 4 , 8 7 1

6 4 , 2 3 4
− 4 1 , 7 8 5
	7 3 2 5 , 0 0 8
	    − 1 2 , 3 4 5
		12 1 5 2 , 1 2 3
		    − 1 1 , 4 7 8

9 6 , 3 6 2
− 2 4 , 5 8 9
	8 9 9 7 , 2 4 6
	    − 8 4 3 , 0 7 2
		13 2 , 3 1 7
		    − 1 , 2 4 9

4 8 , 2 0 0
− 1 2 , 7 3 1
	9 8 9 , 7 3 2
	    − 6 4 , 2 9 6
		14 7 8 , 0 5 1
		    − 1 , 3 8 0

8 6 7 , 2 3 8
− 4 2 3 , 8 5 3
	10 3 , 7 4 2 , 1 7 4
	    − 1 , 6 7 6 , 8 4 2
		15 7 0 , 3 1 3
		    − 2 2 7

There are 80 separate subtractions in Exercise 4.
No more than 6 mistakes are allowed, getting 6 or less mistakes equates to 93% or more. If a mistake is made, it should be explained to the pupil why they went wrong. If more than 6 mistakes are made, they must repeat the whole exercise at a future date until they make 6 mistakes or less.

Example

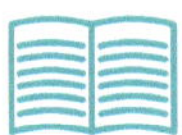

FIND:
$$\begin{array}{r} 3\ 6,\ 4\ 7\ 1 \\ -\ \ 9,\ 8\ 4\ 3 \end{array}$$

3 is greater than 1 so we start using BSM. 3 from 10 is 7, ADD 1 to get 8.

$$\begin{array}{r} 3\ 6,\ 4\ 7\ 1 \\ -\ \ 9,\ 8\ 4\ 3 \\ \hline 8 \end{array}$$

7 is greater than 4 so we drop 1.

7 − 4 = 3, drop 1 to get 2

$$\begin{array}{r} 3\ 6,\ 4\ 7\ 1 \\ -\ \ 9,\ 8\ 4\ 3 \\ \hline 2\ 8 \end{array}$$

8 is greater than 4 so we start using BSM again, 8 from 10 is 2, ADD 4 to get 6

$$\begin{array}{r} 3\ 6,\ 4\ 7\ 1 \\ -\ \ 9,\ 8\ 4\ 3 \\ \hline 6\ 2\ 8 \end{array}$$

9 from 9 = 0, 0 + 6 = 6

$$\begin{array}{r} 3\ 6,\ 4\ 7\ 1 \\ -\ \ 9,\ 8\ 4\ 3 \\ \hline 6,\ 6\ 2\ 8 \end{array}$$

3 drop 1 = 2

$$\begin{array}{r} 3\ 6,\ 4\ 7\ 1 \\ -\ \ 9,\ 8\ 4\ 3 \\ \hline 2\ 6,\ 6\ 2\ 8 \end{array}$$

____

____

FIND:
$$5,256,237 - 1,715,952$$

$7 - 2 = 5$

```
  5, 2 5 6, 2 3 7
− 1, 7 1 5, 9 5 2
                5
```

5 is greater than 3, so we start using BSM.
5 from 10 is 5, ADD 3 to get 8

```
  5, 2 5 6, 2 3 7
− 1, 7 1 5, 9 5 2
              8 5
```

9 from 9 is 0, ADD 2 to get 2

```
  5, 2 5 6, 2 3 7
− 1, 7 1 5, 9 5 2
            2 8 5
```

6 is greater than 5 so we come out of BSM and drop 1.
6 − 5 = 1, drop 1 to get 0

```
  5, 2 5 6, 2 3 7
− 1, 7 1 5, 9 5 2
         0, 2 8 5
```

$5 - 1 = 4$

```
  5, 2 5 6, 2 3 7
− 1, 7 1 5, 9 5 2
       4 0, 2 8 5
```

7 is greater than 2 so we start BSM again.
7 from 10 is 3, ADD 2 to get 5

```
  5, 2 5 6, 2 3 7
− 1, 7 1 5, 9 5 2
     5 4 0, 2 8 5
```

We are now at the end of the line and so we drop 1
5 − 1 = 4, drop 1 to get 3

```
  5, 2 5 6, 2 3 7
− 1, 7 1 5, 9 5 2
  3, 5 4 0, 2 8 5
```

Going in and out and in again is the hardest part of BSM. Students who sail through the previous 4 Exercises stumble on this one.

That is why I am making the following recommendation. Would the teacher please do the first 3 questions with the pupils, consolidating the procedure. As each further question is attempted by the pupil, it should be checked immediately. If an error is made, the error must be pointed out and corrected before the next question is attempted. The pupil should be left to do questions 6 to 15 on their own.

They are allowed at most 4 mistakes out of the 52 separate subtractions in these 10 questions. This equates to 92% or more. If they make more than 4 mistakes, they must repeat all 15 questions again, at a future date, until they make 4 mistakes or less.

1	6 5 4 , 3 7 1 − 1 2 7 , 7 3 8	2	9 4 8 , 2 3 4 − 4 1 9 , 9 2 7	3	2 1 3 , 6 6 3 − 5 1 , 8 8 1
4	7 2 4 , 9 1 2 − 1 4 5 , 8 2 3	5	2 6 5 , 1 2 6 − 2 4 5 , 6 1 9	6	2 1 7 , 8 2 7 − 8 2 , 9 4 1
7	2 0 0 , 9 6 8 − 1 4 7 , 3 8 1	8	2 , 8 1 5 − 1 , 2 8 2	9	7 3 , 0 2 1 − 2 5 , 8 1 0
10	5 2 , 7 8 4 − 3 8 , 2 9 1	11	2 0 , 4 2 5 , 6 2 3 − 4 8 1 , 3 5 7	12	1 7 , 8 6 3 − 9 , 8 2 4
13	5 , 6 7 0 − 2 , 6 7 5	14	1 0 , 1 5 3 − 1 5 4	15	7 0 3 , 2 4 5 − 3 , 2 4 8

You may have noticed, that when you are in BSM and the 2 digits in any column are the same, the answer is always 9 (see 13, 14, 15 in Exercise 5).

You can use this fact to work out the answer more quickly in such cases.

Example 

FIND:
$$
\begin{array}{r}
8,756,434 \\
-\ \ \ 756,484 \\
\end{array}
$$

4 - 4 = 0

$$
\begin{array}{r}
8,756,434 \\
-\ \ \ 756,484 \\
\hline
0 \\
\end{array}
$$

8 is greater than 3 so we start using BSM. 8 from 10 = 2 ADD 3 to get 5

$$
\begin{array}{r}
8,756,434 \\
-\ \ \ 756,484 \\
\hline
50 \\
\end{array}
$$

You are in BSM,
The numbers in each column are the same so we get 9's

$$
\begin{array}{r}
8,756,434 \\
-\ \ \ 756,484 \\
\hline
999,950 \\
\end{array}
$$

Now drop 1 , 8 – 1 = 7

$$
\begin{array}{r}
8,756,434 \\
-\ \ \ 756,484 \\
\hline
7,999,950 \\
\end{array}
$$

If you forget to come out of BSM when the top digit is greater than the bottom digit, you will STILL get the right answer, but you will get a carry figure.
So, for those pupils who do not like going in and out and in again, this is an alternative method.

Example

FIND:

$$36,471 - 8,843$$

3 from 10 is 7 ADD 1 to get 8	3 6, 4 7 **1** − 8, 8 4 **3** **8**
7 is greater than 4 and we should drop 1, but we forget and carry on in BSM. 4 from 9 = 5, 5 + 7 = 12	3 6, 4 **7** 1 − 8, 8 **4** 3 ₁**2** 8
8 from 9 is 1 ADD 4 = 5 + carry 1 = 6	3 6, **4** 7 1 − 8, **8** 4 3 **6** 2 8
8 from 9 is 1, 1 + 6 = 7	3 **6**, 4 7 1 − **8**, 8 4 3 **7**, 6 2 8
3 drop 1 = 2	**3** 6, 4 7 1 − 8, 8 4 3 **2** 7, 6 2 8

In the next exercise (Exercise 6), I will not make any recommendation regarding how many mistakes are allowed. In my experience of teaching BSM, at this stage the student understands the procedure very well and makes very few mistakes or none at all. I leave it to the teacher's judgement as to how many mistakes are allowed and whether or not the student should repeat the whole exercise. The pupil should aim for 95% or more. This equates to getting no more than 5 mistakes.

1.
```
   47,938
 −14,304
 ________
```

2.
```
   87,302
 −47,201
 ________
```

3.
```
   17
 −  5
 ____
```

4.
```
   16
 −  9
 ____
```

5.
```
   4,305
 −1,859
 ______
```

6.
```
   73,578
 −67,879
 ________
```

7.
```
   346,525
 −246,527
 _________
```

8.
```
   368,219
 −164,289
 _________
```

9.
```
   256
 −192
 ____
```

10.
```
   48,391
 −28,294
 ________
```

11.
```
  3 0 0, 0 0 0
− 1 2 4, 5 3 8
```

12.
```
  1, 0 0 4, 6 3 8
−       8, 6 3 9
```

13.
```
  4 0, 4 7 2
−      4 7 1
```

14.
```
  7 0, 4 3 2
−      5 8 7
```

15.
```
  3 2 8, 4 9 2
− 1 6 3, 5 2 8
```

16.
```
  4 3 9, 5 8 7
− 3 7 4, 7 8 9
```

17.
```
  5 6 3, 8 2 9
− 3 4 7, 5 0 4
```

18.
```
  1 0 0, 0 2 3
−        5 8
```

19.
```
  1 0, 0 0 0
−      3 8 7
```

20.
```
  7 6, 9 4 2
− 6 9, 9 6 5
```


100

The following statements are equivalent:

9 minus 7

•

Take 7 from 9

•

Subtract 7 from 9

•

What is the difference between 7 and 9

•

What needs to be added to 7 to make 9

•

How much more than 7 is 9

Example 

How much more than 240 is 496?

We have to work out:
$$\begin{array}{r} 496 \\ -\ 240 \\ \hline \end{array}$$

This is an easy subtraction sum, since the top digits are all greater than the corresponding bottom digits.
The answer is 256.

1 Find the difference between £892.28 and £476.43

2 What needs to be added to 84 to make 136.

3 How much more than £473 is £732.

4 The distance between two towns A and B is 452 miles. If I have travelled 89 miles from A towards B, how far am I from B.

5 Usain Bolt ran the 100 meters in 9.71 seconds at Crystal Palace. His world record is 9.58 seconds. How much slower did he run at Crystal Palace.

6 How much needs to be added to £72.57 to make £100.

7 How much more than 789 is 1,250

8 Take 2,572 from 8,346

9 Subtract £34,000 from £53,505

10 A woman spent £32.56 at a supermarket. How much change should she receive from a £50 note.
(write £50 as £50.00)

Find the missing digit:

```
  4, 3   5
- 1, 8 5 9
___________
  2, 4 4 6
```

In the units column we see that we are in BSM since 5 is less than 9. In the tens column we are either coming out of BSM and dropping 1 or staying in BSM.

If we are dropping 1 the missing digit is 10, which is not a single digit, and so we must still be in BSM.

So we take 5 from 9 = 4 and ADD the missing digit to get 4, and so the missing digit is 0.

Example 

Find the missing digit:

```
  8 6 1
-   6   1
_________
  1 7 0
```

In the units column 1 – 1 = 0, so we are not in BSM.

In the tens column, the answer is 7 which is greater than 6 and so we must be starting in BSM. To get the answer 7 we must take the missing digit from 10 and ADD 6, so when we take the missing digit from 10 we must get 1 since 1 + 6 = 7, so the missing digit is 9.

Find the missing digits.

1
```
   8, 7 6 2
 − 5, 4     1
 ─────────────
   3, 3  3 1
```

2
```
   4, 6 5 3
 − 1,     2 5
 ─────────────
   3, 2 2 8
```

3
```
   5 8, 7 2 1
 − 1 4,     5 8
 ───────────────
   4 4, 0 6 3
```

4
```
   2 0, 0 0 0
 − 1 8,     5 6
 ───────────────
    1, 2 4 4
```

5
```
   3 2, 5 1 3
 − 1 5,     0 8
 ───────────────
   1 6, 9 0 5
```

6
```
   8 5,     3 1
 −    2, 4 7 1
 ───────────────
   8 2, 9 6 0
```

7
```
    4  , 3 2 4
 − 1 2, 5 9 3
 ───────────────
   3 5, 7 3 1
```

8
```
    2, 4     3
 −  1, 5 4 7
 ─────────────
       8 6 6
```

9
```
   4 0,     2 1
 −        5    7
 ───────────────
   3 9, 9 9 4
```

10
```
    7  , 2 4 3
 − 3 7, 4    5
 ───────────────
   3 4, 8 0 8
```

In this section, I will show you my method for doing mental calculations. There are many different ways of doing mental calculations, so if you have a different way that suits you, THEN STICK WITH IT. Do not feel forced to adopt my method. The main thing is that in mental work we strive for ACCURACY, not SPEED.

In the exercises in this section, you show no working, you simply write down the answer, ANY METHOD that delivers the correct answer in a reasonable length of time is acceptable.

We are going to restrict ourselves to examples in the practical situation of mentally calculating the change due from 50p, £1.00, £5.00, £10.00 and £20.00.

The thing to remember is that whereas in written calculations we work from right to left, in mental calculations we work from **left to right.**

So we reverse Bramwell's rule and get
"DROP 1, THEN ALL FROM 9, LAST FROM 10."

Change from 50p

FIND MENTALLY

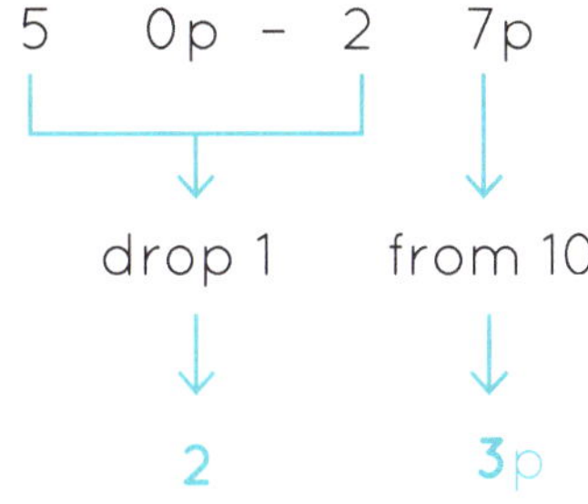

FIND MENTALLY

1 | 50 p – 36p =

2 | 50 p – 22p =

3 | 50 p – 17p =

4 | 50 p – 13p =

5 | 50 p – 29p =

6 | 50 p – 37p =

7 | 50 p – 8p =

8 | 50 p – 20p =

Change from £1.00

FIND MENTALLY   £ 1 . 0 0   –   £ 0 . 4 3

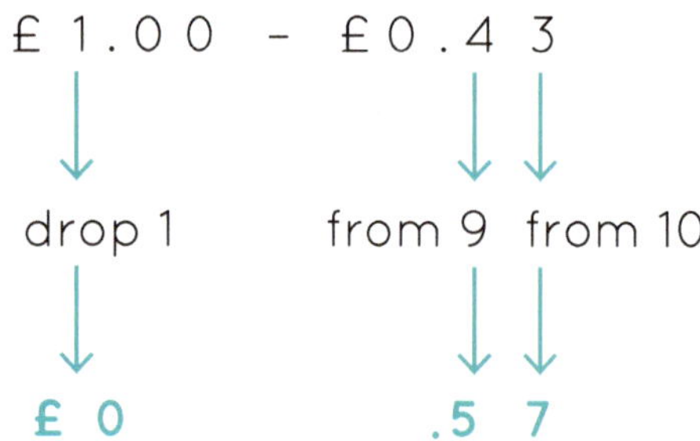

FIND MENTALLY   £ 1 . 0 0   –   £ 0 . 2 7

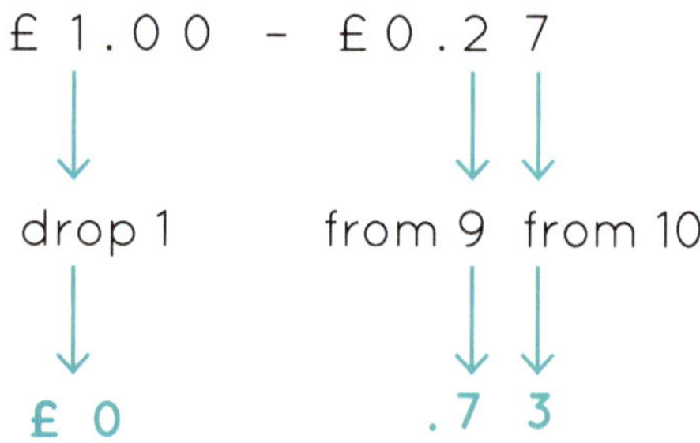

FIND MENTALLY

1 | £1.00 – £0.32 =

2 | £1.00 – £0.72 =

3 | £1.00 – £0.67 =

4 | £1.00 – £0.13 =

5 | £1.00 – £0.59 =

6 | £1.00 – £0.47 =

7 | £1.00 – £0.08 =

8 | £1.00 – £0.30 =

9 | £1.00 – £0.28 =

10 | £1.00 – £0.19 =

Example

FIND MENTALLY

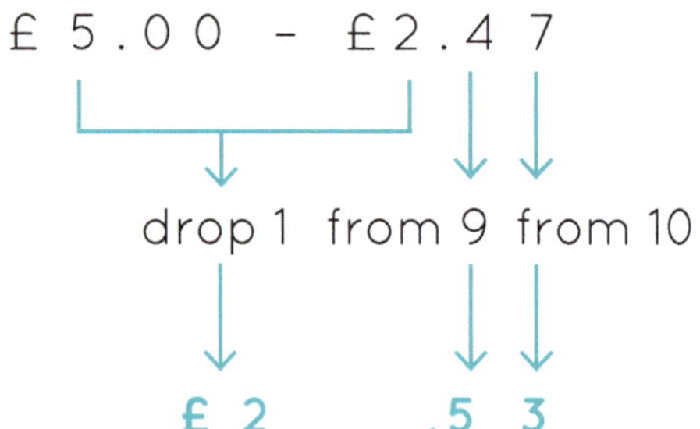

Example

FIND MENTALLY   £ 5 . 0 0 – £ 1 . 2 0

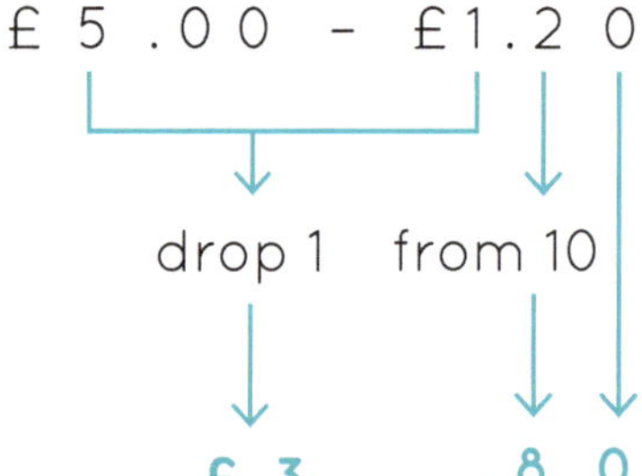

Note that last from 10 means the
last non zero digit from 10.

FIND MENTALLY

1 | £5.00 – £2.37 =

2 | £5.00 – £3.08 =

3 | £5.00 – £1.76 =

4 | £5.00 – £3.40 =

5 | £5.00 – £1.66 =

6 | £5.00 – £4.27 =

7 | £5.00 – £3.56 =

8 | £5.00 – £2.71 =

9 | £5.00 – £3.07 =

10 | £5.00 – £1.70 =

Change from £10.00

FIND MENTALLY

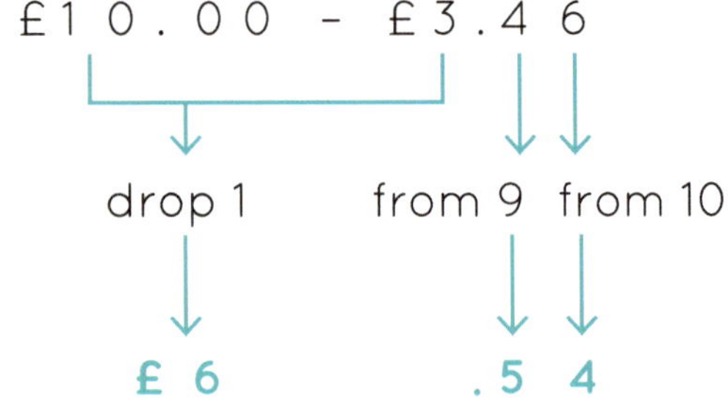

FIND MENTALLY

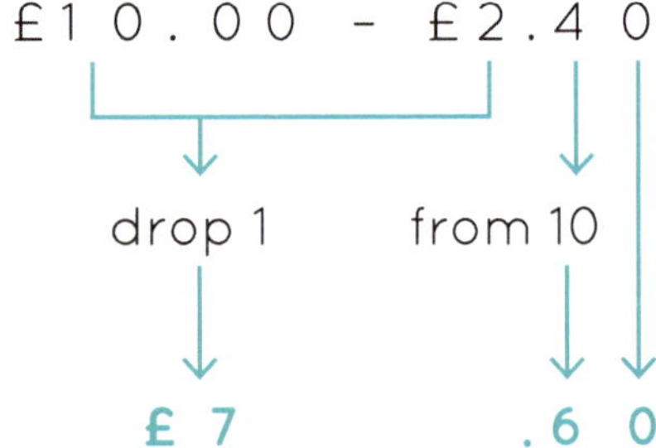

FIND MENTALLY

1   £10.00 – £2.53 =

2   £10.00 – £6.27 =

3   £10.00 – £1.87 =

4   £10.00 – £9.43 =

5   £10.00 – £3.46 =

6   £10.00 – £2.70 =

7   £10.00 – £7.40 =

8   £10.00 – £5.55 =

9   £10.00 – £4.23 =

10   £10.00 – £0.71 =

Change from £20.00

FIND MENTALLY

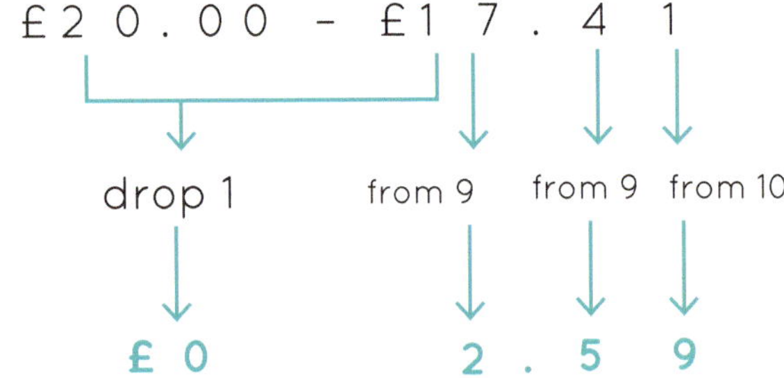

FIND MENTALLY

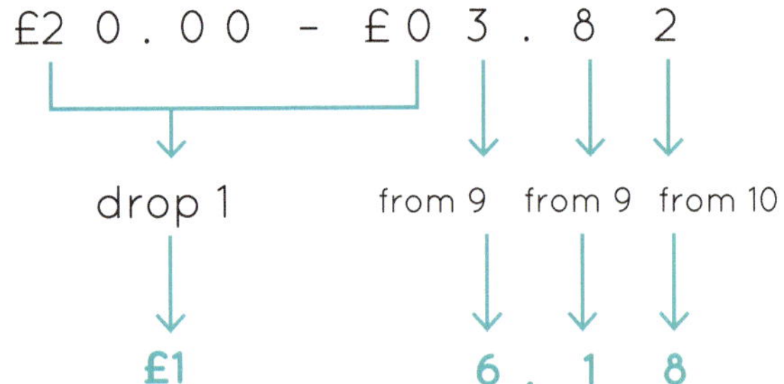

FIND MENTALLY

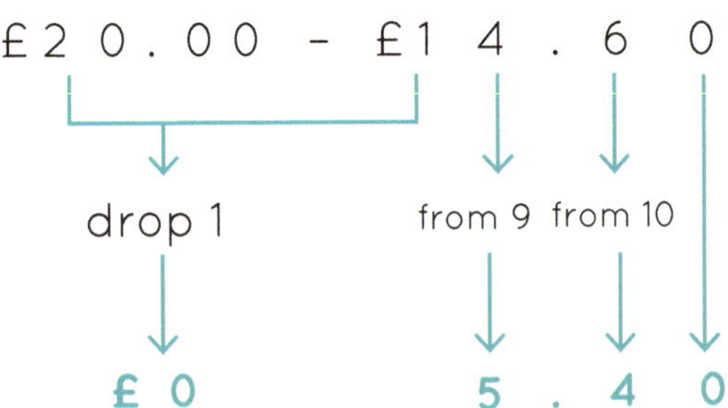

FIND MENTALLY

£ 20. 00 – £ 1 4.3 3 =

£ 20.00 – £ 0 2.8 7 =

£ 20 .00 – £ 0 8.3 2 =

£ 20 .00 – £ 1 6.2 9 =

£ 20.00 – £ 1 1. 40 =

£ 20.00 – £ 0 2.70 =

£ 20.00 – £ 0 3.60 =

£ 20.00 – £ 1 2.6 8 =

£ 20. 00 – £ 1 5.45 =

£ 20.00 – £ 0 8.0 0 =

DR DENNIS L BRAMWELL

**How Bramwell's subtraction
method came to be discovered.**

In 1994, I came across a book
entitled "Vedic Mathematics".
Amonst the many wonderful
things in the book was a quick
way to subtract a number from
100,1,000,10,000, 100,000,.......
This is equivalent to finding the
complement* of a number.

Whilst I was on holiday in
Jamaica in 1996, I found an old
Mathematics book. In it was
a section that dealt with the
subtraction of binary numbers.
This started me thinking about
extending the method to
ordinary numbers. Eventually I
realised that what they did with
binary numbers was equivalent
to turning the subtraction of
ordinary numbers into addition
using complements.

I was very excited about my
discovery and arranged to give
lectures on the new method,
The beauty of the method
was that it did not involve any
borrowing.

My first lecture was at a
primary school in Harbour
View. I asked the teacher to
pick out the weakest student
in Arithmetic. She picked out
a girl who was 8 years old. I
took her aside and taught her
my new subtraction method.
I started the lecture with a
subtraction contest between
the teacher and the pupil.
I made sure that the sum
involved a lot of borrowing.
To my surprise, the pupil froze
at the start, and I thought that
this was going to be a disaster,
but after about 4 seconds,
much to my relief, she began
and finished at the same
time as the teacher. It was
clear that the answers were
different and lo and behold, the
teacher had made a mistake.
The rest of the pupils were
eager to learn my method, and
they listened attentively as I
explained it to them all.

* see Appendix 3, page 58

I was keen to publish my new
method, but the justification
was a bit complicated to
explain. It wasn't until 2008
that I discovered a beautiful
justification of my method that
took just one line.

TRADITIONAL METHODS OF SUBTRACTION

The first traditional method we shall discuss is the decomposition method, also known as the regrouping method or the partitioning method.

THE DECOMPOSITION METHOD

Example 

$$\begin{array}{c} 704 \\ -389 \\ \hline \end{array} = \begin{array}{c} 700 + 0 + 4 \\ 300 + 80 + 9 \\ \hline \end{array} = \begin{array}{c} 600 + 100 + 4 \\ 300 + 80 + 9 \\ \hline \end{array} = \begin{array}{c} 600 + 90 + 14 \\ 300 + 80 + 9 \\ \hline 300 + 10 + 5 \end{array}$$

$$= 315$$

Example 

$$\begin{array}{c} 1{,}000 \\ -\ 256 \\ \hline \end{array} = \begin{array}{c} 900 + 90 + 10 \\ 200 + 50 + 6 \\ \hline 700 + 40 + 4 \end{array} = 744$$

1.
$$\begin{array}{r} 7\ 4 \\ -\ 3\ 8 \end{array} =$$

2.
$$\begin{array}{r} 4\ 2\ 5 \\ -\ \ \ 4\ 8 \end{array} =$$

3.
$$\begin{array}{r} 4\ 0\ 7 \\ -\ 1\ 4\ 8 \end{array} =$$

4.
$$\begin{array}{r} 6\ 1\ 0 \\ -\ 4\ 7\ 6 \end{array} =$$

5.
$$\begin{array}{r} 2{,}0\ 0\ 0 \\ -\ \ \ \ 8\ 5\ 9 \end{array} =$$

Answers
1. 36   2. 377   3. 259   4. 134   5. 1,141

The second traditional method of subtraction is a shorthand version of the decomposition method, called the borrowing method.

THE BORROWING METHOD

Example

$$
\begin{array}{r} 7\ 0\ 4 \\ -\ 3\ 8\ 9 \end{array}
=
\begin{array}{r} ^6\!\!\not7\ ^9\!\!\not0\ ^1\!4 \\ 3\ 8\ 9 \\ \hline 3\ 1\ 5 \end{array}
$$

Example

$$
\begin{array}{r} 7\ 2\ 5 \\ -\ 2\ 6\ 9 \end{array}
=
\begin{array}{r} ^6\!\!\not7\ ^{11}\!\!\not2\ ^15 \\ -\ 2\ 6\ 9 \\ \hline 4\ 5\ 6 \end{array}
$$

Example

$$
\begin{array}{r} 1,0\ 0\ 0 \\ -\ \ \ 8\ 4\ 7 \end{array}
=
\begin{array}{r} \not1\ ^9\!\!\not0\ ^9\!\!\not0\ ^1\!0 \\ -\ \ \ 8\ 4\ 7 \\ \hline 1\ 5\ 3 \end{array}
$$

```
   5 8 8
 -   5 9
 ________
```

```
   8 5 8
 - 4 7 3
 ________
```

```
   3, 2 8 4
 - 1, 6 2 9
 __________
```

```
   6 1 0
 - 4 7 6
 ________
```

```
   2, 0 0 0
 -    8 5 9
 __________
```

Answers
1. 529   2. 385   3. 1,655   4. 134   5. 1,141

The third traditional method of subtraction is the compensation method.

THE COMPENSATION METHOD

$$
\begin{array}{l}
732 \\
-286
\end{array}
=
\begin{array}{l}
700 + 30 + 2 \\
-200 + 80 + 6
\end{array}
=
\begin{array}{l}
700 + 30 +12 \\
-200 + 90 + 6
\end{array}
$$

(We added 10 to both numbers)

$$
=
\begin{array}{l}
700 +130 +12 \\
-300 + 90 + 6 \\
\hline
400 + 40 + 6
\end{array}
= 446
$$

(We added 100 to both numbers)

$$
\begin{array}{l}
304 \\
- 186
\end{array}
=
\begin{array}{l}
300 + 0 + 4 \\
-100 + 80 + 6
\end{array}
=
\begin{array}{l}
300 + 0 +14 \\
-100 + 90 + 6
\end{array}
$$

$$
=
\begin{array}{l}
300 +100 +14 \\
-200 + 90 + 6 \\
\hline
100 + 10 + 8
\end{array}
= 118
$$

1.
$$
\begin{array}{r}
4\ 2 \\
-\ 1\ 8 \\
\hline
\end{array}
$$

2.
$$
\begin{array}{r}
3\ 6 \\
-\ 1\ 9 \\
\hline
\end{array}
$$

3.
$$
\begin{array}{r}
5\ 8\ 8 \\
-\ \ \ 5\ 9 \\
\hline
\end{array}
$$

4.
$$
\begin{array}{r}
8\ 5\ 8 \\
-\ 4\ 7\ 3 \\
\hline
\end{array}
$$

5.
$$
\begin{array}{r}
6\ 1\ 0 \\
-\ 4\ 7\ 6 \\
\hline
\end{array}
$$

Answers
1. 24   2. 17   3. 529   4. 385   5. 134

The fourth traditional method is a shorthand version of the Compensation method, called the Austrian method.

THE AUSTRIAN METHOD

$$\begin{array}{r} 7\ 2 \\ -\ 4\ 8 \\ \hline \end{array} = \begin{array}{r} 7\ 2 \\ -\ 4\ {}^{1}8 \\ \hline 2\ 4 \end{array}$$

The little 1 serves two purposes. It indicates that we are subtracting 8 from 12 in the units column, and also that in the tens columns, we are adding 1 to the 4 so that you are taking 5 from 7. This is equivalent to adding 10 to both numbers in the compensation method.

$$\begin{array}{r} 6\ 0\ 4 \\ -\ 4\ 8\ 8 \\ \hline \end{array} = \begin{array}{r} 6\ {}_{1}0\ {}_{1}4 \\ -\ 4\ 8\ 8 \\ \hline 1\ 1\ 6 \end{array}$$

1.
```
   8 4
 - 3 7
 ______
```

2.
```
   3 6 5
 - 1 2 8
 ________
```

3.
```
   3 6 5
 - 1 8 8
 ________
```

4.
```
   5, 2 0 4
 - 2, 8 9 6
 __________
```

5.
```
   4, 1 5 1
 - 2, 7 5 5
 __________
```

Answers
1. 47   2. 237   3. 177   4. 2,308   5. 1,396

WHY BRAMWELL'S SUBTRACTION METHOD WORKS

First we have to define **complements.**

If two numbers less than N, add up to N, they are said to be complements (base N), where N is a power of 10. (10, 100, 1,000, 10,000,.............)

eg. 600 and 400 are complements (base 1,000). because 600 + 400 = 1,000.

The complement of a number A (base 1,000) where A < 1,000, is 1,000 - A

Example

There is a very quick way to find the complement of a number.

Find the complement of 6,580 (base 10,000).

We simply take all the figures from 9 and the last non zero figure from 10

6 from 9 is 3
5 from 9 is 4
8 from 10 is 2
so the complement is 3,420.
(10,000 – 6,580 = 3,420).

And we see that subtracting from 100, 1,000 , 10,000,......is very easy indeed.

The rule is
"All from 9 last from 10"**
The justification for this is easy,

CONSIDER

$$\begin{array}{r} 10,000 \\ -\ 2,146 \\ \hline \end{array}$$

If we subtract using the borrowing method (page 52) we obtain

$$\begin{array}{r} 9,99^{1}0 \\ -\ 2,14\ 6 \\ \hline \end{array}$$

and you can see that you are taking all from 9 and the last from 10.

** See Vedic Mathematics by Sri Bharati Krsna Tirthaji Page 12

$a - b = a + (N - b) - N$, where N is the relevant base number (100, 1,000,....)

To express this in words, because we start from the right, "first from 10, the rest from 9" is simply "all from 9, last from 10" in reverse and so we are actually finding the complement of the bottom number and adding it to the top number.

Example

In the example 704 – 389
we work out:
$704 + (1,000 - 389) = 704 + 611$

$$\begin{array}{r} 704 \\ -389 \end{array} = \begin{array}{r} 704+ \\ 611 \\ \hline 1,315 \end{array}$$

and drop the 1 at the end means get rid of the 1 in the 1000's column which is equivalent to subtracting 1000 to give the answer 315 and clearly

$704 - 389 = 704 + (1,000 - 389) - 1,000.$

We refine this in the last column (hundreds column), instead of taking 3 from

9 and adding 7 to give 13 then dropping the 1, to get the answer 315, we simply take 3 from the 7 = 4 and drop 1 to give 3.
The justification for this is as follows

$$\begin{array}{l} A \,\ldots\ldots\ldots\ldots \quad \text{where } A > B \\ -\ B \,\ldots\ldots\ldots\ldots \end{array}$$

In the first method we take B from 9 and add A to get $9 - B + A$, then drop 1

which is $9 - B + A - 10$

In the refined method we calculate $A - B - 1$, which is the same thing.

EXERCISE 1

1. 241	6. 1,431
2. 231	7. 5,325
3. 711	8. 1,121
4. 241	9. 2,002
5. 496	10. 5,463

EXERCISE 2A

1. 258	6. 6,658
2. 1,158	7. 7,137
3. 1,222	8. 19,358
4. 6,364	9. 29,266
5. 3,472	10. 39,836

EXERCISE 2B

1. 8	6. 1,875
2. 8	7. 672
3. 17	8. 2,238
4. 283,775	9. 4,255
5. 8,484	10. £22.38

EXERCISE 3

1. 3,833	6. 251	11. 94
2. 27,568	7. 185	12. 39,783
3. 38,923	8. 573	13. 9,780
4. 3,541	9. 2,522	14. 5,523
5. 334	10. 2,705	15. 94,730

EXERCISE 4

1. 22,176	6. 261,786	11. 23,852
2. 22,449	7. 312,663	12. 140,645
3. 71,773	8. 154,174	13. 1,068
4. 35,469	9. 25,436	14. 76,671
5. 443,385	10. 2,065,332	15. 70,086

EXERCISE 5

1. 526,633	6. 134,886	11. 19,944,266
2. 528,307	7. 53,587	12. 8,039
3. 161,782	8. 1,533	13. 2,995
4. 579,089	9. 47,211	14. 9,999
5. 19,507	10. 14,493	15. 699,997

MIXED EXERCISES 6

1. 33,633	6. 5,699	11. 175,462	16. 64,798
2. 40,101	7. 99,998	12. 995,999	17. 216,325
3. 12	8. 203,930	13. 40,001	18. 99,965
4. 7	9. 64	14. 69,845	19. 9,613
5. 2,446	10. 20,097	15. 164,964	20. 6,977

EXERCISE 7 (WORD PROBLEMS)

1. £415.85
2. 52
3. £259
4. 363
5. 0.13 seconds

6. £27.42
7. 461
8. 5,774
9. 19,505
10. £17.44.

EXERCISE 8

1. 3
2. 4
3. 6
4. 7
5. 6

6. 4
7. 8
8. 1
9. 5 at the top,2 at the bottom
10. 2 at the top, 3 at the bottom

EXERCISE 9.

1. 14p
2. 28p
3. 33p
4. 37p

5. 21 p
6. 13 p
7. 42 p
8. 30 p

EXERCISE 10

1. 68p
2. 29p
3. 37p
4. 83p
5. 44p

6. 57p
7. 93p
8. 62p
9. 72p
10. 81 p

EXERCISE 11

1. £2.63
2. £1.92
3. £3.24
4. £1.60
5. £3.34

6. 73p
7. £1.44
8. £2.29
9. £1.93
10. £3.30

EXERCISE 12

1. £7.47
2. £3.73
3. £8.13
4. 57p
5. £6.54

6. £7.30
7. £2.60
8. £4.45
9. £5.77
10. £9.29

EXERCISE 13

1. £ 5.67
2. £ 17.13
3. £ 11.68
4. £ 3.71
5. £ 8.60

6. £ 17.30
7. £16.40
8. £ 7.32
9. £4.55
10. £ 12.00

NOTES

NOTES

NOTES